Calculus

Mastering the Mathematical Language of Change

(Practice Workbook With Worked Examples and Practice Problems)

Edward Hicks

Published By **Regina Loviusher**

Edward Hicks

Calculus: Mastering the Mathematical Language of Change (Practice Workbook With Worked Examples and Practice Problems)

ISBN   978-1-7775767-7-6

No part of this guidebook shall be reproduced in any form without permission in writing from the publisher except in the case of brief quotations embodied in critical articles or reviews.

Legal & Disclaimer

The information contained in this book is not designed to replace or take the place of any form of medicine or professional medical advice. The information in this book has been provided for educational & entertainment purposes only.

The information contained in this book has been compiled from sources deemed reliable, and it is accurate to the best of the Author's knowledge; however, the Author cannot guarantee its accuracy and validity and cannot be held liable for any errors or omissions. Changes are periodically made to this book. You must consult your doctor or get professional medical advice before using any of the suggested remedies, techniques, or information in this book.

Table Of Contents

Chapter 1: A General Method For Solving 1

Chapter 2: Adding Consecutive Odd
Numbers 23

Chapter 3: Converting Line Intuition Into
An Analytical Form 47

Chapter 4: Functions Of Variables 59

Chapter 5: Combining Functions 72

Chapter 6: Basic Tools Graphs 102

Chapter 7: Algebraic Function 105

Chapter 8: The Chain Rule 115

Chapter 9: Product Rule 121

Chapter 10: The Quotient Rule 125

Chapter 11: Trigonometric Functions... 129

Chapter 12: Exponential And Logarithmic
Functions 131

Chapter 13: Derivative At A Point 138

Chapter 14: Second Order Derivatives . 141

Chapter 15: Integration 145

Chapter 16: Definite Integrals 152

Chapter 17: Rules Of Domain Of A Function 157

Chapter 18: Limit Tending To A Real Number 166

Chapter 19: Derivatives Of Trigonometric Functions 174

Chapter 20: Velocity And Acceleration. 178

Chapter 21: Basic Standard Integral 181

Chapter 1: A General Method For Solving

Mathematics Problems

Most people become lost in math because they think of the subject as a complete set of equations and formulas that must be learned.

Yet, remembering formulas and equations is just scratching the surface of math.

There is no reason to be surprised that a person who has written down such information find math boring and tedious.

In this chapter, I'll describe the plan to a generic method of finding more challenging mathematical problems.

If one's experience is just having a list of formulas and then applying them can be quite a shock to come up with a broad way to solve questions. To be clear the method doesn't provide you with the solutions. Its purpose is help you prepare your mind to find the answers. If the problem is straightforward and you are familiar with the formula, this isn't a good idea. If it's an extremely difficult issue and you're not sure how to proceed the formula can aid you. Though we will not refer to these specific steps throughout the text, I believe that you'll find these ideas in the course of we explore the principles behind the calculus.

2.1The Process

2.1.1Step 0: Write it Down

Prior to moving on to the actual actions, make certain to record the issue. It may sound easy, however many tend to ignore this step due to it being to be repetitive. Write down the issue is a good way to ensure you are aware of what your problem asks. I almost failed in my statistics course at college because I would always answer wrong questions and omitting key aspects of the question. Recording the question in the way the question is asked will prevent your mind from taking numerous mental shortcuts as well as preventing your from overlooking crucial information which will aid you in solving the issue. By writing the question down, you can ensure that the problem itself is carried across your head to reach your hands.

2.1.2Step 1: Find Out What the Problem Means

I love to say I like to tell people that "step 1 of math is philosophy." What does that mean is that issues are not only numbers that we can plug into an equation we were given by someone else. They only function when we have seven

2. A GENERAL METHOD FOR SOLVING MATHEMATICS PROBLEMS

definitions and, the most important initial step is understanding what's meant by the query. 1

It is the task of the philosopher. Philosophy is concerned with the issue of significance. It attempts to understand an unambiguous wording to make it definitive. If you have the opportunity to go through the writings of

Plato. There's no reason to believe that Plato's theories are perfect, or even excellent. The main persona in his writings, Socrates, is able to help those he interacts with in making their concepts more precise, precise and rational through asking pertinent inquiries about the meanings and significance of the words they choose to use.

Socrates always made sure that his audience for them to clarify the meanings of the terms they were using. He posed tough questions on these concepts to make the person he was speaking with to comprehend all specifics and meanings of the words they were referring to in order to make a better and more specific use of their own language as well as to determine if there was unclearness or inconsistency with their terminology.

The same is true for mathematics. If you are faced with a challenge The first thing to do is to understand what you're discussing. Ask yourself questions regarding the significance of the issue so to know the issue you're seeking to discover. If your understanding of the issue is ambiguous or contradictory, it will be difficult in locating the solution. If you are able to formulate a exact meaning of the issue this way, finding an answer becomes much more straightforward.

In order to determine the root of a problem There are two broad types I prefer to employ two categories: ignorance and knowledge. If you want to master mathematics effectively, you have to be able to be particular about your skills as well as your lack of knowledge. Consider Algebra. In Algebra it is very precise about the lack of knowledge we have, marking the area by a"x. If I tell you, "three times some number is twelve" then I could convert it into a mathematical equation, marking my lack of knowledge ("some numbers") by marking it with an x. Then, I can say 3x=12. Since I have clearly marked my ignorance I then can solve this equation, and then say that x = 4.

It is possible that we are ignorant of several areas. In this case, we can simply write more letters (x, y Z, A, B C, etc.) Each one expressing the specific lack of knowledge. If you try to confuse things which you're not aware of and you do the mathematical incorrect.

Most of the time, we often have more than one identical amount. They are usually distinguished by subscripts. In the example above, if I discuss the distance Bob and Jill have each traveled and I want to give each distance a distinct alphabet, like the letter x to represent Bob's distance, as well as y to represent Jill's or choose a single letter for distance and then use the subscripts for each. If I do that it would be the letter d Bob to represent Bob's distance and for Jill's distance, d Jill to represent Jill's distance.

Sometimes, mathematicians reduce their subscripts to make d B corresponds to Bob's distance while d J represents Jill's distance.

They can also attempt to abstract Bob or Jill and substitute them with numbers. For instance, that d_0 corresponds to Bob's distance while D_1 represents Jill's distance. 2

In all these situations in all of these cases, we're being clear about the lack of understanding, employing variables to show what we don't know in mathematical equations.

Alongside being exact in our knowledge, knowing the root of the issue requires being specific about our understanding. This involves several factors. In the first place, we must note any connections that we can discover from the actual problem as well as try to record them in formulas in the event that the issue is particular enough to allow that. Then, we need to note any additional information which we have gathered about the subject.

1It's a common misconception that the science of physics is developed mostly via experiment. This isn't to say that experiments aren't essential, but if you are a scientist, it's unlikely that what you are doing all day long is an experiment after another. There are some who do, but the main tool used by the philosopher is in fact the philosophy. Einstein created his equations when being a patent officer. The patent clerks aren't given extravagant labs or the chance to use any lab at all. The equations for relativity just by considering what it means when energy and matter were the same. What would the world look like If this was true? For the most part, in science nowadays, philosophy has been all just a memory. However, philosophy is all about getting your mind to think clear on difficult questions that could be presented in ambiguous ways. Philosophical methods help us be able to think more clearly even when other people think mindlessly.

2Also take note that mathematicians begin counting from zero, while others begin counting at the point of 1. Thus, one mathematician could employ d0 and 1 while another may be using both d1 and.

2.1. THE PROCESS

When we're in the realm of triangles, what details we should know? It is true that all angles be 180 degrees (or P Radians). When the triangular shape is right triangle it is known that we could utilize to solve the pythagorean theorem. If the triangle isn't one of the right triangles, we are able to draw a line to convert the singular triangle in two right triangles which make it easier to manage. Also, we know the formula to determine the area of a triangle and diverse rules regarding triangles like those applicable to similar triangles.

When we're dealing with circles, we have an additional set of facts. We are familiar with the equations that determine both the circumference and the area. It is possible that you have that formula for the length of an arc in the circle. It is possible that you have the formula to calculate a graph of circles.

When faced with any issue in any situation, it's best to write down every fact that is mentioned in the issue. If you're stuck, occasionally just taking a look at all the details which you have gathered which are connected to the relationships and facts identified in the problem can be beneficial. Reviewing them can help to discover connections and connections you've never imagined previously.

The process of learning and applying math not just about numbers as much as it's about learning facts about the world which are related to each the other. Formulas, numbers, and equations are just the tools that enable us to achieve this.

The most beneficial type of relation to understand for mathematics is equations. These ought to be written like equations when feasible. However, even if it's not possible to understand how to write the relationship in equations At the very minimum, write down the relationships you've learned about.

2.1.3

Step 2: Converting questions into Answers

What I find the most remarkable about math is the fact that it lets us convert exact questions into precise answers through the inner logic of the query itself. Take a look at this. When we are faced with an issue, we need to look for answers. You have to go through an article, research the world, or talk to somebody. In mathematics, our goal is to formulate questions in a clear enough manner that the query is transformed into an solution.

This can be a valuable technique for the rest of your lifetime. That means that many aspects we are thinking about, once we've learned to be exact in what we think about The question itself will tell our what we need to know. Mathematical reasoning is just a way to practice the kind of thinking we do. 3

For converting a query to a numerically fixed answer (i.e. that we need to determine the exact amount of the number x) We must at a minimum, have an equal number of equations we use to calculate not know the value. It is not always necessary to have an exact answer, sometimes we're simply trying to find an interesting relationship. Whatever the case an equation's number will give us the largest amount of variables that could be diminished. It isn't always possible to use equations to lower the number of variables you have to count. Sometimes, two equations can tell that you exactly the same thing which is why they do not have to count as separate equations. At most it is possible to limit the number of variables by the maximum amount of equations to work with.

It is often possible to convert inquiries into solutions by changing the words. For instance, suppose we have two equations below:

The equation $y = x + 5$

$x y = 6$

We have here the equality of y with $5 + x$, which implies that wherever I find the word y, I could replace the y with 3A fantastic book that provides an abundance of useful tips for adding more precision to your questions is How to Measure Anything: Understanding the value in "Intangibles" in Business by Douglas Hubbard. The author demonstrates that almost everything that is abstract it is, could be turned into a quantifiable quantity.

10

2. A GENERAL METHOD FOR SOLVING MATHEMATICS PROBLEMS

5 + x. If we replace y (five + x) within the 2nd equation we find a new equation (x + 5) = 6(5 + x) = 6

Be aware that it's recommended when making substitutions, to place the substitution in parentheses at the time of first doing the substitution, to make it obvious that the sequence of operations will be properly applied. Since y was only a one unit in the initial equation, what that it is replacing becomes a single unit. the parentheses will ensure that it is applied in the correct way.

If we reduce this equation, we get the following equation

5. x 2 + = 6

It looks like something. What could it be?

It looks almost like a quadratic equation.

Indeed, if

When you tweak it just a little and it transforms into an equation that is quadratic:

5 + 2 = + 6 = zero

Now it is possible to go back to the first step and find out what equations mean. If we've got the quadratic equation, do we are aware of any specific formulas that relate to it? Actually, we could apply the quadratic formula to find the answer (actually an answer set comprising two solutions) in a direct manner:4

- b +- b 2 - 4 ac

"x" =

2 a

-5 +- 52 - 4 * 1 * (-6)

"x" =

2 * 1

-5 +- 25 + 24

"x" =

2

-5 +- 49

"x" =

2

-5 +- 7

"x" =

2

Therefore, x could be one or the other.

In other words, even if be converted into responses, often what we do is to convert the questions we have asked into questions that we need to return to the first step and consider repeatedly.

2.1.4

Step 3: Convert Our Steps Into a Formula

It is not always needed, but it is the reason math can be beneficial. A single solution to a problem can be beneficial.

The ability to solve a problem in the way that you can are able to formulate a solution that you can use for the rest of your life is more beneficial. When you are in life (whether within business, in the sciences or construction) there are times when you have to deal with similar quantitative issues repeatedly.

They could be solved each time. Better yet could you create the formula to ensure that you only need to solve the problem once and you never need to do it ever again. After that, it's just plug and drink your results.

In order to convert a sequence of steps to a formula the first thing you must accomplish is 1. Find out what parts of your formula will remain fixed and which ones will change between time to time when you apply it.

If you're not familiar with the quadratic formula, then you must review it as well as other formulas included in appendix H. This quadratic equation is utilized frequently throughout this book.

2.2. APPLYING THE METHOD

11

2. Replace those parts that change every time (but remain constant with respect to a certain application) with a variable that will symbolize these parts. These are not considered variables as such since they don't change in the context of the issue. In fact, I refer to them as constant variables since in some ways, they behave like variables (they differ with each time you apply the formula) However, inside their equations, they behave

as constants. This distinction might not have been a major factor in your math professional life before, however, as we'll see coming up the calculus constants and variables are regarded differently.

3. Then, follow exactly the same steps as the previous steps, only by using the constant variables you created to substitute for real figures.

4. In the end you'll have an algorithm for repeating the exact same procedure repeatedly.

If you're not sure regarding this process the process will be clearer through the example that follow in the following section.

2.2

Applying the Method

In this part in this section, we'll look at several applications of the approach to problems that are simple. These are issues that can be solved easily enough with no help However, having them implemented with simple issues will to get your brain on the right track to utilize the method to solve more complex questions.

2.2.1

Bob and Jill's Travels

Bob and Jill depart from the same point that is in opposing direction. Both travel in a different, yet constant speed for 3 hours. Bob is travelling at 25 miles an hour while Jill is at 30 miles per hour. In 3 hours of travel, how much from each other are they?

The first step is to determine what exactly the problem is:

1. What do we want to find? We're trying to find the distance that they covered. This distance will be called"x" and name the thing it is that we're not aware of.

2. Bob and Jill depart from a point. In another way, it implies that their distance from the beginning is equal to zero.

3. Bob and Jill have been traveling in opposing directions. If they were travelling with a different angle to one another, it could be a challenge However, since they're traveling in opposite directions, this means the distance between them is the distance they traveled each.

4. Because we do not know the exact distances Jill and Bob traversed, we'll depict their travels with additional variables- the variables d J and D B.

5. Bob and Jill travel at a high speed. What's a speed? Speed refers to distance .

time

6. What's the relation between distance and speed? Distance simply means speed multiplied by time it takes to travel.

7. What was their travel time? In the question, it gives an solution of three hours.

8. What speed are they travelling? The answer is provided with the following

question: Bob is travelling at 25 miles an hour and Jill is at 30 miles per hour.

It is possible to convert a number of them into equations. You can transform 1 2, 3, and 4 into a straightforward equation such as the equation: x = d B + D J

12

2. A GENERAL METHOD FOR SOLVING MATHEMATICS PROBLEMS

We're not yet aware of the d J and d B however that's fine. These points 5,6 8, and 7 permit us to create equations for miles

D B = 25

* 3hrs = 75 miles

Hour

miles

D J = 30

*3 hours equals 90 miles

Hour

The second step is to transform our query into an answer by using the logic that is embedded in the question. There are three unanswered questions (x, dJ and the d B) however, there are also three equations (two of which we have just have to solve). Thus, we are able to put our solutions into the equation that we have to solve and obtain what we need:

Chapter 2: Adding Consecutive Odd Numbers

What three odd numbers sum to 111?

To resolve this issue, we'll need to accomplish a few tasks. One is to acknowledge our insecurities. We don't know anything. We aren't familiar with all three numbers. Therefore, we'll give each number an individual letters to convey our insecurities: the letters x, y and the letter z. Now that we've declared our ignorance in detail this problem statement lets us to transform the following equation:

$x + y + z = 111$ 2.2. APPLYING THE METHOD

13

This is a total of three unsolved questions, but just one equation. For solving the problem there must be at minimum three equations. But, we are not completed our inquiry into the significance of the question.

The issue states, "consecutive odd integers." What are we aware of about odd numbers? In reality, half the numbers are odd. Particularly, every other number is strange.

Furthermore, they're consecutive meaning one following the other. How do you go from one odd number to the following one? Simply add two! Thus, if we choose the first number as x it means that we could move to the next number (y) through addition of two to the number x. You can record that in the form of an equation

The equation $x = 2$

What is the z? As it's also an consecutive odd number and we are able to tell that it needs at least two times more than one. This is enough to be an equation, too.

$z = y + 2$

There are three equations that represent three variables. It is time to move onto 2nd step which is making our problem into a solution. It is possible to take our original equation, and then substitute z for by adding x and y. $(2 + y) = 1111$.

Then we could substitute the letter y:

(x + (2 + x) + (2 + (2 + x)) = 111

Rearranging and regrouping a part, we are able to transform the result into:

*x + x + 2 + + 2 equals 111

3 x + 6 = 111

We are now at a point of solving this problem:

3 x + 6 = 111

3 x = 111 - 6

3 x = 105

The x value is 105.

3

35 x 35

We can figure out that y is 37, and the z value is 39.

This is why Step 3 is even easier as the sole constant that has to be changed into a variable constant is the value that is finalized. It is possible to use the variable v to accomplish this. We can then return and perform an equivalent transformation: the formula x + y + Z = the variable v.

$x + y + (2 + y) = v$

$(x + (2 + x) + (2 + (2 + x))) = v$

$3 x + 6 = v$

We now have an equation which can be used to solve a similar issue. If I ask "what three consecutive odd integers add up to 237" then you could simply add that for the initial answer (x) 3x + 6 = 237

$3 x = 231$

$* x = 77$

14

2. A GENERAL METHOD FOR SOLVING MATHEMATICS PROBLEMS

A side note: this assumption assumes that the value can actually be obtained through the addition of three odd numbers. If your results are an even number, or a decimal, it cannot be obtained with the help of three consecutive odd numbers.

2.3

Additional Tips

A lot of students struggle to know what to do if they are stuck or having difficult time finding the right answers. These tips aren't comprehensive, but they can aid you in keeping the tips in your mind while working on issues.

2.3.1

Don't Solve the Problem Immediately

In some cases, even for the most basic questions, you are able to resolve your issue in a matter of minutes. This isn't never a good idea for dealing with concepts that are new. One of the problems when you have to solve a problem quickly is that you've failed to take the time to fully understand the problem.

Remember that the first step of mathematics is philosophy. Make sure you take the time to comprehend the issue prior to attempting to resolve it.

2.3.2

Don't Skip Steps

When my children are struggling with math most of the time it's due to missing the steps. Being impatient and striving to avoid steps is harmful both ways. First, it is more likely to commit a mistake. By writing down every process, you become more mindful of how you're doing it as well as allowing you to quickly spot your errors. Furthermore, other individuals will help you identify the mistakes you could make. Over time solving problems, these steps can help to embed in your brain how to solve the issue. In the beginning, skipping steps, especially at the start of your beginning to learn, will prevent the pattern of the issue from becoming permanently etched in your brain.

2.3.3

Draw the Problem

If the problem is anything which is conceivably drawable and you can draw it, then complete it. Since you're contemplating the issue in a visual and an analytical way It will enable you to engage more of your brain to solve the issue. Once you have a clear picture of the issue, often the answers or how to solve it is more obvious.

2.3.4

Focus on the Next Step

Don't expect to resolve the issue all in one go. There is a possibility of alternating back and forth in between steps 1 and 2 as often as you want. Do one step and then take a look at the problem and again. Is the problem different today? Are there avenues that are now open? Do you want to press ahead or return to scratching the surface?

If you do not force yourself to tackle the entire task all at once, rather move on to the next task that is in front of you, frequently you will find a solution which you'd never thought of prior to.

2.4. CONCLUSION

15

2.3.5

Ask "What If "

One of the most important aspects in solving problems is to ask "what if" questions. For trigonometry or geometry, one might ask, "what if we drew a line here?" What could that mean for how we approach the issue?

What would happen if we draw the line to create an right triangle?

Also, think about equivalents that you are aware of. Sine and cosine have a relationship. What happens when you change an equation of sine to a cosine function - will it bring you closer to the solution?

Sometimes, you come across an equation or a problem that resembles an earlier problem you've encountered. It is then possible to ask whether there is something I can add to solve this issue that will make it simpler to resolve?

In the same way as we're trained to try to decrease the number of variables we have to consider, the addition of one variable can make an issue more like the problem that we have to deal with.

As you begin to learn math, the ability to formulate these kinds of questions is not as strong. But as you study increasing amounts of mathematics and become more proficient, your capacity to ask the right "what if" questions increases. Trigonometry and geometry is the ideal way to begin this sort of thinking.

An excellent example is within Section 4.7.

2.3.6

Think Through a Simplified Example

When you're finding it difficult to solve a problem you can look at a simplified model of the issue. What's that is making it difficult? How can you resolve the problem without a hurdle?

If yes, how do you change your mind to incorporate the issue?

The simple elements is the way we can to ultimately find the answers to difficult questions.

2.4

Conclusion

A lot of people believe philosophers are very separated from maths, as it is a hard science, as philosophers stare at their nails. This notion is not true. Philosophy, in its essence offers the fundamental techniques of thinking. Certain of them are deeply embedded in our minds that we can't see the world with a different lens. That doesn't mean philosophical thought isn't useful, it is just that it has become so important that we can't imagine our lives without it.

Philosophical thinking asks two kinds of inquiries: questions about what is to be authentic, and the ways we know whether it is real. Philosophical thinking allows us to construct mathematical theories by gaining the significance of every word. To build equations, we need to understand the meaning of a concept in the most fundamental levels. This is called the philosophy.

A mathematical equation is essentially nothing more than a equivalency of two concepts. 5. To determine that two concepts are comparable is to be able to comprehend the philosophical meaning behind them. Without this understanding, we would not be able to determine if they are equal. When we are certain that they're identical, we can apply the mathematical principles (which are in turn derived from the philosophy of mind) to alter them, and discover additional truths.

In my classes almost every lesson I teach students I require them to respond to philosophical question first: What do they mean by the question? It is a way to avoid using formulas that aren't applicable, and in 5Technically, it can be any relation and not only equality. Equations are also equations. The general principle is applicable.

16

2. A GENERAL METHOD FOR SOLVING MATHEMATICS PROBLEMS

is of helpful to you at any time to come back. Being able to tackle the question of meaning before anything else will help you think for your own self. In the real world, you frequently need to trust experts. Experts can be great, but of time, the questions they answer aren't identical to the ones that you're asking. Learning to understand the meanings behind your questions will let you know what answers you're receiving are in line with what that you're asking. It will also help determine whether the expert has an appropriate solution to your issue.

3

Basic Tools: Lines

If you are planning to venture to the woods it is essential to be sure your bag is equipped with the necessary equipment you'll need. In the first chapter, you can be sure that the essential concepts from past years are on the top of your list as you read this guide.

3.1

The Line

A simple line can be a extremely powerful mathematical tool. Lines are straight path in space (whether in three, two or additional dimensions) which extends forever across all directions. When we speak of the term "line" in this book the term is used to mean the straight line. Curved lines will be called the term "curve. The term "line" may refer just to a portion of the line between two points, but it is better known as an a segment of a line.

Many theorems and formulas and equations are founded on the characteristics of lines and calculus is not an exception. So, reviewing lines is crucial.

In all potential lines you can draw within a graph for determining an exact line, there are only two points on the line. There is no difference if you're working in just two, one, or 3 dimensions (or greater) Any two points can be used to define the definition of a line. So long as two points do not happen to be identical, it is possible to draw only one line linking and including the two points. The line doesn't end at the points you have chosen however, it extends forever in all directions. These two points serve only to help you find just one of the many lines which you might be thinking of. Two distinct lines will serve as the basis for defining the line you are looking for.

3.2

The Equation of a Line

Lines in two dimensions are drawn an easy, common equation:

The equation y = mx +

(3.1)

The equation is composed of four parts:

It is a number which corresponds to the horizontal portion of the graph of a line.

17

18

3. BASIC TOOLS: LINES

It is a term that is related to the vertical part of the graph of a line.

It is a constant which represents how steep the lines slope. The slope is the amount of units are added to the graph for each unit it moves towards the right. It is commonly called "rise over run" (i.e."rise over run" or r ise).

r u n

It is a variable which indicates how far the line shifts towards or away from its beginning (a positive value indicates that the line is moved up and a number that is negative indicates that it's been shifted down while zero indicates that it crosses the line of origin).

Here's a graph of the equation: $y = 3 \, 1 + x$:

7

6

5

4

3

2

|

1

0

-1

-2

-3

-4

-5

-2 -1 0 1 2

x

If you look at the equation and compare each of the terms to its corresponding values in the conventional Line Equation (Equation 3.1) If you do this, you'll see that m = 3. b is 1. B tells you the location at which the line is going to intersect the y-axis. Because b equals 1, and the graph is then shifted by 1 unit away from the beginning, meaning that it intersects the y-axis in 1. Verify the graph for that.

In the present, the slope (m) is 3. That means the line will climb three units for each unit it is able to go right. When the slope is positively it will be slanting to the right. If it is negative, the line will tilt towards the left. Lastly, in the case of zero slope, the line will remain smooth.

Slopes can be represented using ratios, even if they do not have to be. This is to assist users realize that the slope is due to a ratio of shifts in the y direction as well as the change in the direction of x. For this example it is the case that the slope is 3 that, when represented as a ratio is . If 3 units are added, the value of y change, the x 1.

Value changes by 1. The slope may also represent a number it can be read as the value is rising by more units for each unit to the right is moved.

Let's take another look:

$$y + 1 = 1. (x + 3)$$

2

It's not exactly the traditional line equation nevertheless, it is an equation. To make it match with Equation 3.1 it is necessary alter it a more. First, you need is to spread one of the coefficients over the (x + 3) phrase. This will result in two

3.3. DETERMINING A LINE FROM TWO POINTS

19

$y + 1 = 1. x + 3$

2

2

The second is the closest, but there's a constant that is in the left-hand aspect of the equation, which is required to be moved towards the right to be able to find Equation 3.1. Thus, we must subtract 1 (which also means the number 2) from each side of the 2 equation, yielding:

$1x = 1x + 5$

2

2

This equation is now it being the same equation as the one of. The slope is 1 which means there are 2 equations for each

One unit of change in the y value is changing 2 units of value of x. The value of b is set at 5, which implies that it will change 2

The y-axis is intercepted at -2.5. Get a sheet of graph paper. See whether you are able to draw the line based on the data provided.

3.3

Determining a Line from Two Points

In the past, we mentioned that a line could be drawn from any two points. You can easily think of this using a graph paper that has two dots across it. If we use an appropriate ruler, there's only one method to join the dots to form an unidirectional line. Additionally to that, when we create the line straight, we have numerous possible points along this line. If we chose the two points of the line, using the ruler to join these two points would result in exactly the same straight line.

When we take a look to the graph papers, and then think about how we can join two points in the form of a line, we're using our intuition and thinking geometrically. This means that we are applying our common sense to understand the things we are seeing. But, it's important to understand mathematics in order translate our ideas into a format that is easier to comprehend and be able to compute with. It would be wonderful to take the notion that lines are created from two points, and then convert it into a formula which can be converted from two points into equations to create an equation for a line.

Formulas such as this, that are based on intuitive concepts and transform the ideas into algebraic format which we then utilize to manipulate and calculate and calculation, are known as analytic form (also called closed forms) due to the fact that they offer an accurate method for understanding the issue with higher precision and less intuition. Analytic forms let us not just make our thoughts crystal clear and precise and precise, but also create formulas that can allow us to make our thoughts more useful and teaching-able. Furthermore, analytical forms are easy to incorporate using a computer to automate future tasks.

In this instance, can be we take the notion that a line can be created from two points and transform it into an equation that is an equation? What is the best way to do this?

If you go back to Equation 3.1 The term "m" is used to describe the slope. What exactly is an incline? What exactly does the word mean? Referring to the prior article, a slope can be described as one of the ratios between changes in y direction to the changes in the direction of x. The term is commonly used to mean "rise over run", or"rise over run," or . As this is a proportion that means "ru nu

regardless of what distances between any two lines are, the percentage of changes is identical regardless of the points you choose. So, if there are at least two points, you are able to determine their slope using the ratio between those differences between the x and y values of each point.

Let's suppose that there are two points two points: P0 and 1. P 0 is situated in the location (2 3, 3) and P 1 is situated at location (4 and 6). In order to determine the slope, it is enough to figure an equation of the change in y and the changes in the value of x. So, given two instances such as P 0 and 1. The slope between these two is:

20

3. BASIC TOOLS: LINES

The formula m = y1-y0

(3.2)

1x1 x 0.

This equation states that y0 represents the coordinate for y of P 0. and x0 is the coordinate for x of P 0. and y1 is the location of the P1 and x1 is the coordinate for the x of P 1.

If we are using the above points this is:

M = 6 + 3

4 - 2

This provides

M = 32

For every increase of 3 our graph creates the graph goes above two.

The equation now is just half completed. In the end it is:

Chapter 3: Converting Line Intuition Into An Analytical Form

Further, we could offer a comprehensive analytical equation for the line in relation to the two points.

It is generally simpler to work out the equation for an equation like we did in the preceding section. But having an Analytical solution to the equation can be beneficial in many situations. In the case of example, if we were to create a computer program that calculates the equation for two lines and a completely analytical version of the equation would be beneficial since computers are not able to make sense completely.

To figure out how to develop this concept, there is already an equation to solve for the variable m (Equation 3.2). We just have to rewrite the line equation in order to find the equation b:b = y + the mx (3.3)

As we've used a certain value (P0) We can replace the x and y values. The result is an equation: b = y0 - the mx value 0.

In the equation for that slope will get:

B = y0 + 1 x 0

(3.4)

1x1 - x0

Therefore, by combining Equations 3.1, 3.2, and 3.4 We can then combine them in this massive equation the formula: y = y1 - the sum of y0 + y0 = the y1 equation: y1 + y0 x 0.

(3.5)

1x1 x 0x0

1 - x 0.

This is a completely analytic problem, as you will be able to discover the solution simply by substituting the formula and taking the calculated. There is no need for any logical thinking, simply put in the plug and go. But, if you were to apply a formula such as this but it's not going to help your comprehension of how various parts work together.

The ultimate goal of math is usually to find an analytical form for the equation. But, when learning math the most important thing is to develop an intuitive understanding of how different parts connect to one another, as well as, in the second place, understanding how to create an analytical equation using the most intuitions.

Equation 3.6 is the shortened version that is a shorter version of Equation 3.5. Equation 3.6 is typically found in textbooks. However, the purpose in Equation 3.5 consisted of teach the user how to develop an equation on your own by using the fundamentals of what lines are.

$y - y0 = y1 - y0 (x - x 0)$

(3.6)

$x 1 - x 0.$

To test your skills to test your knowledge, try changing the layout of Equation 3.5 so as so that it appears like equation3.6.

22

3. BASIC TOOLS: LINES

--

Example 3.1

Based on the two points (7 and 2) and (2 5, 7) Determine the equation for the line between them.

In order to solve this question, we can apply either a step at an time (solving the equation for m and then finding the b) or solve it with Equation 3.5 as well as Equation 3.6. Let's use Equation 3.5. Start with equation:

y = y1 - y0 x + y0 - y1 - y0 x 0

1x1 x 0x0

1 x x x 0.

Then, we'll swap the two points. The point doesn't matter if is used as points 0 as well as Point 1 insofar that we're consistently. If we use (7, 2) for P 0 and (7 2, 7) as P 0, the 7 to represent x0, and 2 for the y0.

So, (2 5) will result in P 1. therefore we'll use 2 to represent x1 and 5 for y1.

In the process of substituting them into our formula, we will get:

y = y1 - y0 x + y0 - y1 - y0 x 0

x 1 -x 0

1x1 - x0

y = 5 - 2 x + 2 - 5 - 2 7

2 - 7

2 - 7

y = 3

- x + 2 - 3 7

5

-5

-3

-21

y =

x + 10 +

5

5

5

-3

y =

The sum of x + 31

5

5

By using the equation above, we observe how the slope of this equation is 3 and the y-intercept 31 .

5

5

When you consider the ruler as an example that you will get the exact lines if you draw the line starting from the initial point and ending at the second point just as you do if drawing starting from the 2nd point towards the point at which you started.

It is likely that you have played with lines in a few ways during your previous Algebra classes. Slopes and lines, even though it may appear like they are simple, are crucial for the calculus. Thus, it is important to make certain that you are able to comprehend every aspect of this prior to moving on to the next chapter.

Review

We learnt:

1. A straight line is a route through space in the 1, 2 3 or even n dimensions.

2. A line can be completely defined by the two lines or traced using a ruler to join both points.

3. An infinite line is made up of points.

4. The most common version of the equation for the line is that $y = mx +$ in which m is the slope and b the intersection of y.

EXERCISES

23

5. It's simple to draw lines if they are drawn in the regular format.

6. With two factors (which create the lines) It is simple to determine the equation for the line they define.

7. The combination of all our knowledge of lines enables us to create an analytical explanation of the method to calculate lines from point coordinates which is where you only need to do is enter points coordinates, and it pours the equation that results.

Exercises

1. Create the equation for one line (Equation 3.1) Five times.

2. What is the meaning of m in the equation that defines an equation?

3. What is the meaning of b on the formula for the line?

4. Draw the line using the equation: $y = 5\ 3\ x$.

5. Draw the line that is given by the equation $y = x + 1$.

3

6. Draw the line in the equation the equation $y = 2.3\ x + 4.1$.

7. Draw the line that is given by the formula $5y = 3\ x + 10$.

8. Draw the line that is given by the equation $y + x = 2$.

9. Calculate the equation for the line defined by the angle (2 1, 2) as well as the slope 1 .

5

10. Calculate the equation of the line drawn by numbers (1 1) and (2 3,).

11. Find the equation for the line drawn by two points (4 5,) and (-3 2,).

12. Provide an analytic formula to calculate the equation for the line when given only one place (both x 0 and) as well as the slope (m).

24

3. BASIC TOOLS: LINES

4

Basic Tools: Variables and Functions

This introduces an array of concepts and lots of terms. Pay particular concentration on the concepts which are discussed, since they are crucial to comprehension of concepts to come later.

4.1

Variable Relationships

When you study advanced math it is necessary to deal with functions and variables. A variable can be described as a placeholder for an amount. In some cases, a variable represents undetermined value which needs to be solved. In the example above, if I carry an amount of money in my pockets There is just one number for x although we do not know the exact value.

Sometimes variables represent the possibility of a range of values which can be utilized. Let's take for instance you have an investment plan that will increase the amount of money you invest. Then, we can construct an equation where the x is the sum of money that you put into the investment while y is how much money you'll receive after the time period of investment. As the investment doubles in value when you invest it, we can express this as the equation $y = 2x$

In this instance the x number doesn't refer to an exact number of numbers that you can choose from any quantity you'd like to. It only informs that you, on the basis of the amount you invest (x) what amount of money you'll receive (y). This is why you can call x an independent variable since you are able to decide how much you put into it. The amount you receive (the sum of money that you get) is considered to be dependent due to the fact that its value is dependent on the amount you put into it. Variables are referred to as co-dependent when their value is dependent on the other.

There isn't any justification for using specific x and/or y numbers for these numbers aside from the tradition of customary. Mathematicians frequently utilize letters to represent numbers because they're simpler to write into equations. If you're comfortable using these names for variables using shorter, standard names for variables makes equations simpler to understand. When solving equations with two variables that are independent, the variable typically utilizes the letter x, and when graphed, it is positioned horizontally. The dependent variable generally utilizes the letter y, and is a vertical axis.

Chapter 4: Functions Of Variables

If we consider the equation that y = 2x we can say that y is the explicit expression of the value of x. It means that y's entirely dependent upon the value that is given by the x equation, and not dependent on any other variables. Functions can be described as a device that receives any number of values and generates a result (usually the only one) in the form of an output.

If two x = y it is possible to say that the y function is x since the sole item on the left side of the equation the value of y. Everything on the right side of the equation is the function of the value of x (meaning that everything on the right side is dependent on its value in contrast to everything else).

Functions aren't equations. They're merely elements of equations. The function can take up the entire side of the equation. If one of the sides of the equation has one variable and another side doesn't have the same variable, then we could claim that the single variable is a result of variables that are located on the opposite side the equation. The other aspect of the equation could be described as a computer, which in the event of being supplied with the values of the other variables creates the single variable.

Before we get into calculation, the majority of graphs you've observed are actually functional, and you should take some time to contemplate what means for something to be not functional. Take, for example the circle's graph:

It is important to note that for virtually all values of x there exist two possible value of the value of y! The value of y's worth isn't only based on the values of the x. In order for y to be able to hold an exact value, you need to understand the significance of y! In this instance the variables x and dependent variables. That is, neither is completely independent of the one. The circle is not the result of a function because there was a possibility that for every valid x there is multiple possible values of the y. To summarize it is possible to say that y could be described by a function that is explicit to x when you consider that for each possible value of x there is only one number of values for the y.

If you have a graph An easy method to find out if the graph is an expression of x, or not is by using the test of vertical lines.

A vertical line check declares that if any vertical line could be drawn that would meet with the graph more than one time The graph is not an accurate dependent variable.

Sometimes, you can work around this problem by choosing particular numbers over other. In this case, for instance, we could find y

4.3. IMPLICIT FUNCTIONS

27

such as:

9 9

Equation for a circle with diameter 3

$9 = y2 \times 2$

Find the y on its own

$y = 9 - x2$

Both sides of the square root

The square root can have two value, negative and positive. Therefore it's not really considered a function. We can still consider this an actual function by choosing to only take one of the square root. How an appropriate choice depends on the situation in hand.

In addition, you may consider expressions that produce different values as an unsuitable method.

This means that it could perform the same way as the function at any moment to allow us to assume it's an actual function in the majority of cases. Also when you zoom in to the graph in a way that is close then the part that you zoomed in on the graph will appear to be function. This lets you treat it as a function many cases.

4.3

Implicit Functions

In some instances in which y could be the result of x it is possible that the function may be defined in a way that is implicit rather than explicit. Take a glance at the following formula:

x y = y + x

In this formula, the two variables xs and ys are merged to form the equation. However, this doesn't mean the equation can't be written in an equation. In this instance, you definitely could (y is x). Sometimes, however, the x-1

finding a variable such that it's solely on one side can be difficult and shouldn't have to be accomplished. When this happens, the problems can be treated in the form of implicit functions. A implicit function is one that can be described as an expression of x however, the equation doesn't segregate the ys from the parts of x. Thus, we can see by the vertical line test it's the graph of an equation, however it is necessary to perform more research to find out what the function actually is.

4.4

Naming Functions

Variables and functions may be identified with names. If there's just one particular function to be considered and it has a name, the usual one is an acronym like f . Similar to when we have many variables, they are called x, y and z. Names of multi-functions are usually known as f , g, and the latter. Sometime, they're given long, specific names (think about cos and sin) However, generally when it comes to mathematics, single letters dominating discussions.

Functions are identified by an e-mail list of parameters, a list of variables function uses in order to calculate its values. This means that the case with f (x) signifies that the function f is able to decide its value using only one parameter, the value for the parameter x. Function called g(x and the y) implies that the function will calculate its value on the basis of two variables, both x and. In these instances it is said that it is a function of the variable x and that the function g depends on the two parameters x and.

What are the function names utilized?

Function names can be used in a myriad of ways. Sometimes, they're employed to describe an in-depth process just as variable names can be used to denote unknown value. Function names can be used to denote specific pieces or notions of a particular function. You can state "for any function f (x) this is the case ..." and

28

4. BASIC TOOLS: VARIABLES AND FUNCTIONS

Then, we can talk regarding the aspects of the x. When we talk about functions of x, when we use an f (x) then we could mean

"any expression we want, as long as it is a function of x."

Let's assume that I inform me that I've formula for determining the amount of money I need to invest in savings every year, based upon my income. It is assumed that x represents my earnings, while the number y indicates how much funds I invest in savings this year, based upon that income. So far, I've stated that the amount I put aside is determined by my earnings. It's not clear which formula I utilize.

Maybe I'll put 10% in savings. Maybe I put everything above $50,000 into savings. We haven't discussed it with you. What you do know at this present is that it's solely based on my earnings. Also it is clear the sum I save depends on the amount I earn that year. Thus, we can say that you (the amount I save) is dependent on the number x (my salary) but we don't have a clue as to what that function is so we are unable to define the equation. We can however, understand that:

$y = f(x)$

The most common translation is " f of x." This means that there exists a function named f, that is able to take one input, which is called"x. The output of this function, whichever it is will be what we convert into the y.

The aim of every mathematical formula is to communicate, no matter if there's lots or little information to convey.

There is a possibility that we would like to present the entire equation that relates my income to my savings However, we can just know that whatever the calculation is the equation is definable in relation to my income (the x). So, $y = (x)$ offers the reader numerous details. It states that y is indeed dependent on x however, it does not specify what the function actually means. Also it is when we state that the equation y = f (x) we are aware that the only thing the y function is x. There aren't any other variables to take into consideration.

Let's suppose that I say that I'll use the first 10,000 of my earnings to cover expenses. Then from the remainder, I only save the other half. What's the formula to connect these two figures? The amount that is saved is 50percent (0.5) from the total salary (*) after the deduction of $10,000 off.

It means that:

$y = 0.5(x - 10, 000)$

We have previously stated that y is a function of (x) - (x). So, if we substitute the word f (x) instead of y in the formula above, it will mean:

$F (x)$ is $0.5(10000 x) = 0.5()$

Thus, whereas previously, we had only vaguely understood that f was the result of x We now know precisely what is f (x) signifies.

Similar to solving for variables you can also find a solution for functions.

This problem in a minute. However, first take a look at something different: salaries. Let's suppose that I've got another equation that connects my pay to the calendar year. This equation lets me know how much I will earn in relation to the year which is the year in which I first joined in the workforce. In this equation there is an independent variable, which is the year and the dependent variable is amount of salary. In the common sense for calculating y, it will be what is the dependent variable (salary) as well as x is an dependent variable (year).

It is also the case that y is a result of the x. To differentiate this particular function from the preceding one, let us refer to the function as"g"(x). Thus, y =(*x). Let's suppose that I inform you that my income was $15,000 at year zero increasing to $1,000 each year thereafter. The equation is:

4.4. NAMING FUNCTIONS

29

$y = 15000 + 1000\,x$

Therefore,

$G(\,*\,) = 1500 + 1000\,x$

We now have two tasks. F (x) uses x as my income, and returns how much I've saved on my income. G(x) considers the year x, and calculates the size of my pay is, depending on the year in which it's. This can get extremely confusing because for each of these calculations each one, x is a distinct amount. For example, in f (x) the word x is used to refer to my earnings, while in the g(x) it refers to the years.

If you are using functions, the x you refer to by the function will be a locally variable in that the value of it is derived from the parameter that is given to it. This does not mean that it is the exact that x is in the other equations. Thus, for example If I'm given g(3) and g(3), this gives me a number where I will take the number 3, and substitute it for each occurrence of x that is defined in the formula of the function. Therefore, g(3) becomes 15000 + 1000 * 3 or 18000.

In the next scenario, suppose I am aware that Phil relocated into New York the year after I earned $19,000. What can we do to determine which year Phil made the move? We'll say"x" represents the year in which Phil made the move. This means that in the the year x 1 - 1 I earned $19,000. Thus, we could make an equation like:

G(x - 1)) = 19000What does the term g(one x) refer to? It is the meaning that, when we are using the term g(x) then we replace each occurrence of x within the definition of the word g(x) by x - 1.

Chapter 5: Combining Functions

One thing you might have observed about the functions such as f (the x) and the function g(the x) in the earlier scenario is that the function f (x) accepted pay as an independent variable and then gave an amount of savings as dependent variable.

But in the case of the case of(x) the salary will be the dependent variable, whereas the year's number is the independent variable.

Then, we could add a year's date to the g(the x) and then get a pay out, and place a salary in the f (the x) and take an amount of savings out, wouldn't it be possible to develop an algorithm that lets us enter a year and then get the amount that I put into savings?

The process of linking two functions together is known as function composition. The way to write it is as follows:

H(x) = f (g(x))

in which the h(x) is the function we have built by wiring the output of G(the x) to the input of the function f (the x). For a better understanding of how it works we must first extend the function's innermost part that is the g(x). It will result in an equation H(*x) = f (15000 + 1000 x)

We can apply 15000 + 1000 x as an input to the variable f (the x). When we extend the definition of the word f (x) and this gives us:

F (x) is 0.5(10000 x) = 0.5()

H(x) = f (15000 + 1000 x)

h(x) = 0.5((15000 + 1000 x) - 10000)

H(x) is 0.5(1000 1000x + 5000)

H(x) equals 500 x + 2500

We now have an analytic, explicit function that takes in the year's number and then tells me how much amount of money I intend to invest in savings. For example, for the year 11 the formula would be h(11) is 500 *11 + 2500 = 5500 plus 2500, which is 8000.

So, for the year 11, I'd put $8000 into the savings accounts of my account.

4.5. COMBINING FUNCTIONS

31

There's no limit on the amount of functions that could be linked to create new functionalities. Additionally, the process of composing functions isn't the only method to connect two tasks.

We can say for example that I give my Jack $3.50 per hour more to do a job, that I would pay Jim to do the same job.

Thus, if the value f (x) is the sum of money I have to pay Jim for working for x hours and the g(the x) is the sum of money I make to Jack for the x hours of work, I will be able to demonstrate the connection between the two amounts using the formula that (x) = g(x) = f (x) + 3x

Let's say the two members work as team to the same number of hours. If we assume that

(x) is the amount of money that they earn, and h(x) is the sum of money the team receives in x hours (i.e. Jim's salary and Jack's) We can calculate that (x) = h(x) = f (x) + G(x)

Since we have the meaning of G(x) We can conclude that it:

H(x) = f (x) + g(x)

G(x) = f (x) + 3x

H(x) = f (x) + f (x) + 3 x(in x) is 2* (the x) + 3x

As you realize, functions can be put together in various ways. We've only looked at specific instances of function. But, to be able to calculate you'll need to think of the functions abstractly.

Also, in the case above it is easy to see the relationship between money and pay and comprehend the fact that paying two people implies that we have to add the payments of both parties. But, in mathematics, it's necessary to understand how manipulate equations and formulas with no clear and concrete definitions of each word. Abstract manipulations are known as formal manipulations, since they only deal with algebraic formulas, but they do not focus on the fundamental concrete knowledge.

In order to ensure you comprehend the idea Let's take a take a look at an example of a function that is not assigned a concrete significance to ensure that you are aware of the concept.

Let's assume that we serve two roles:

F (*) = 2x + 2x + 3

G(x) = 2x + 3

Let's look at that. the h(x) in which case

H(x) = f (g(x))

If we swap the value inside first, we'll be able to be able to:

H(x) = f (2 x + 3)

If we then take 2 x + 3 as the parameter to calculate the function f (x) and rewrite it, you will end up with:

32

4. BASIC TOOLS: VARIABLES AND FUNCTIONS

H(x) = (2 x + 3)2 + 2(2 3) + 3

In expanding and distributing terms we'll be able to be able to:

H(x) is 4 12 + x + 9 + 4x + 6 +

By combining terms We will have:

H(x) equals 4x2 + 16 18

4.6

Inverse Functions

Inverse functions are the ones is created when you attempt to execute a program in reverse. If, for instance, I run a program that divides by 2, the opposite of that particular function could be one that divides two.

Inverse functions take the output of a certain function and then returns the input.

The inverse function is usually recognized with the addition of a -1 tiny letters beside the name of the function. Therefore, given the function:

F (*) = 2x

The reverse of this function could be described in the form:

F 1 (x) = x 2.

We know, therefore, that whatever value we type for x into the form of f (x) and we'll get this number back when we convert the results of the formula f (the value of x) into the f-1 (the value of x). Also, the reverse function is a mathematical formula that the inverse function for any valid x the formula f 1(* f (x)) is x.

In the above example you can observe that if we add 2 into the number f (x) we'll end up with 4. When we plug 4 into the number f-1(

x) you will receive 2 (our initial number) to return.

Inverse functions may not be as simple, however. If the function that was originally used is more complicated and complex, it may not be simple to determine what the reverse would look like. Methods to find an inverse of a function is quite simple. The first step is to use the function to determine y. the formula. Then, alter the equation using algebra, so that x can be the only one only on the one hand of equation. The opposite side of the equation, in case you swap x for it, will be the result of the function that is inverse.

--

Example 4.1

Let's suppose that the formula f (x) = 2 + x. What's the value of f(*x)?

Then, firstly then, we'll set it equal to y, meaning that 2 = + x. Then, we'll apply algebra to find the x on its own

4.7. INVENTING FUNCTIONS AND THE MAGIC OF SUBSTITUTION

33

On one hand on the other side

$y = 2 e x+1$

Our first equation

$y = e x+1$

divide each side by 2

2

"Y"

Ln

$= x + 1$

Take the log that is natural to each side

2

"Y"

Ln

- 1 = x

Add 1 to each side

2

If x exists on its own, to find the reverse function, we simply take the portion of the equation with the variable y. We substitute in the variable x the variable y. This gives us:

x

F -1 (the x) = ln

- 1

2

A problem with reverse functions is that in the majority of cases, their counterparts are not really a function. It is important to remember that every function generates one value out of the inputs it receives. But, if the graph from your initial function fluctuates both up and down, the function's inverse has multiple results in a specified x.

Consider, for example, the equation that $f(x)$ = $x 2$. It is a real function, meaning that for every x given, there's just one possible result.

What's the opposite? The reverse of this equation is $f\text{-}1(x)$ =

x. But, square roots could be positive or negative

This is a negative number, and therefore it can be better portrayed as 1 (the value of x) = +x. Therefore, for any given number of x values, it could be two possible value.

When the exponents are higher, you can find more possibilities for the inverse function. Certain functions, like cos and sin, are inverses. an infinite number of inverses that can produce outcomes for any given number. In these cases, reverse functions usually have the predetermined limit of possibilities for results.

Most of the time it's not an issue However, it's good to be aware that, even if $f(x)$ can be considered to be a valid function, $f(x)$ may not be or might need been used only with certain requirements.

4.7

Inventing Functions and the Magic of Substitution

In the section 4.5 we examined how functions can be used to analyze them and blend them to create new functions. In this part we'll look at how we can take an equation that is already in place and create new functions that will help us examine equations in new light.

We will look at the following problem:

$\sin 2 (x) + 2* \sin(x) + 1 = 0$

What can be done to solve this problem?

To solve this problem, I would like you to look at different kinds of equations that you've come across throughout your time working with math. Are they reminiscent of something, even though it's not exact? When you look at it, the first word is squared and the second one is in the first power and the final term can be described as a constant. What kind would look like?

If you answered "quadratic equation," you could be right. If not examine the equation closely and see what you could find

34

4. BASIC TOOLS: VARIABLES AND FUNCTIONS

at the very least, you can see the similarity between quadratic equations and the above equation. To remind you that the fundamental form of a quadratic equation

$Ax2 + BX + C = 0$

(4.1)

in which a, b and that c are all constants. For solving this equation for x, apply the quadratic equation. To help you this formula, it can be described as:

$- b +- b 2 - 4 ac$

The x sign is

(4.2)

2 a

There is a issue--the equation isn't quite matching that of the quadratic equation. Every time we're supposed to be able to find an x, instead we get the sin(the x) instead. What do we do?

When you are aware of any place that we would like to have sin(x) that is instead sin(the letter x) We could create a program to transform one of them into another.

What we really wanted was x. But what we received rather was sin(sin = x). So, we'll create the new variable the u.

It is then possible to say:

$U = \sin(x)$

Then, we've determined, through our creative powers we would like to introduce one new variable, u. It should be equivalent to sin(x). In each place there is the sin(x) could be substituted the sin with u, in accordance with the definition we have chosen.

Therefore, replacing sin(x) by u results in the following equation:

u 2 + 2 u + 1 = 0

The equation is now an equation that is quadratic! There is only one difference: it utilizes U instead.

The rules for algebra are the same regardless of how the names of our variables.

So, by using the quadratic formula, we will determine the following:

-2 +- 22 - 4 * 1 * 1

U =

2 * 1

-

= 2 +- 4 - 4

2

-

= 2

2

= -1

Thus, we know that the u value is 1. Do we have it all figured out? Not quite. Keep in mind that the initial issue was finding the x factor, not the an equation that would solve for u. When we completed this transformation, we constructed an equation which ties to x and u. For instance, that u is sin(x). Since they're identical, it is possible to substitute in the opposite way and put sin(x) into the u, and then solve the equation as follows:

U is 1

U = Sin(x)

sin(x) = -1

Since both of them are the same as u, and u, they're equal to each and each

4.8. MULTIPLE VARIABLES

35

In the present, the reverse of action in sin(x) is called arcsin(x) (also called sin-1(x)). When we apply this to both sides of the equation, it would reverse what we did with sin(x) to the left

arcsin(sin(x)) = arcsin(-1)

*x. arcsin(-1)

because arcsin is the reverse of sin, it is possible to obtain an x

x -1.5708

If you get -90, this is because the calculator you are using is using degree mode rather than radion mode. Calculus is a science where we typically compute using radians rather than degrees, unless we have a specific reason.

We'll go back to our initial step-by-step procedure. The first thing we considered was whether our baffling equation could be connected in any way with the more common equation we were already familiar with. After we realized the equation was akin to a quadratic equation we searched for ways to change the equation into a format which we could utilize the quadratic formula to solve. The main issue before us was the formula was using sin(x) instead of x for every term. So, we chose to substitute, substituting for u (which is an expression of the x) instead of sin(x). The result was the shape which we were looking for to work out with the quadratic equation. Once we had solved the equation the equation for u, we required to solve for what we called x. Our original equation was used connecting x and u to determine the answer to u using the number x. We then solved the equation the equation for the x.

This may sound like a number of steps. But it's really just that simple. We came up with a new variable could be used to substitute an additional variable so that it is solvable. Be cautious in doing this, however since you need to ensure that the substitute is able to replace all the x instances in the equation. If not, you've just boosted the quantity of variables that are in the equation. This isn't necessarily a bad thing but, in general, it's unproductive.

In this case, for instance, if you had used sin2 (x) + 2x + 1, this substitution could not be able to work. It would result with u 2 + 1 + x = 0 but it is not solved in any form. The only difference is that we have many more variables.

Making up variables to replace elements in equations is an extremely powerful technique however, you need ensure that you're employing it properly.

4.8

Multiple Variables

Functions can be dependent on several variables. In this case, for instance the distance someone can travel when they are moving is dependent upon their speed as well as the length of time they travel. So, it is possible to create a function that represents the following:

D (speed, time) = time * time

The function of d () (i.e."distance) is determined by two variables: speed as well as time. Thus, in this particular function speed as well as time can be considered independent variables while the distance traveled will be considered the dependent variables.

In most cases, when a function is composed consisting of two variable, x as well as y are the two dependent variables. z will become an independent variable. This can be expressed as graphs with three dimensions. As an example, consider the formula called f (x (x, you) = 3xy. If we were to calculate to calculate the value of f (4 5, 4) then we will find 3 * 4 * 5 equals 60. So, the dependent variable (which is likely to be graphed using the symbol the z) is 60.

It is important to note that some functions contain multiple dependent variables.

36

4. BASIC TOOLS: VARIABLES AND FUNCTIONS

4.9

Logarithms and the Natural Log Function

In the course of your math education it is expected that you are familiar about logarithms as well as what they're employed to do. However, many students are unaware of logarithms until the time they start their calculus. Logarithms are the reverse of exponents. In other words, the Loga (b) determines what I must raise the value to obtain the value b. The a value is regarded as the basis of a logarithm. Thus, log2(8) (which is translated in the form of "log base 2 of 8") is 3. 23 is 8.

Two unique logarithms are the log10(x) and loge (x) in which e represents Euler's number. It is about 2.71828.

If you notice the word log(x) but no bases listed the most likely explanation is the log10(the x). The more crucial, however is the an ln(x) it is an abbreviation of"loge" (the x). Calculus (and elsewhere in many locations) the Ln(the x) is one of the most frequently employed logarithm due to its simplicity in the calculation of many operations in calculus. It is referred to as the natural logarithm due to the fact that it is simpler to understand and use as compared to other logarithms. Calculus (and in general) when there's any reason not to choose an alternative base for a logarithm then the standard logarithm can be a good choice to make.

If you are unable to recall the rules for exponents They are listed in the Appendix H.1.4. One of the key aspects of logarithms are they are able to transform exponentiation into multiplication.

As an example, suppose I have the equation z = y , but the experient causes me to have trouble (for any reason). You can use the natural log on each side, and result in the equation: of ln(Z) = ln(x and). The rules for logarithms in Appendix H.1.4 stipulate that an exponent in the logarithm is relocated outside of the logarithm using an exponent multiplier. Thus, the equation is: ln(Z) = y LN(x). In other words, through the use of logarithms we changed the problem of exponents to one that involves multiplication. It has numerous applications for calculus, and even more.

Review

We have learned:

1. Variables may be used to refer to an individual value or any number of different values which can be employed to solve a problem or function.

2. Independent variables are those variables by which the different variables (dependent variables) are derived from.

It's called "free" because the person who is using the equation typically gets to pick the values. Variables dependent on each other are based on their values in independent variables.

3. Independent variables are typically depicted along the horizontal (x) Axis, while dependent variables usually are plotted along the vertical (y) the axis.

4. Variables can be considered to be codependent if their value is dependent on the other.

5. The majority of the commonly-used variables or function names don't have significance, and they are used because of historical conventions. The conventions are followed, which makes it simpler to comprehend the meaning of what's being described.

6. A function is an array of mathematical operations which produce an amount that is proportional to the value of a variable independent or variables.

7. The term "function" is implicitly defined in the event that it is described by the equation

of two variables x and y, where the dependent variable doesn't sit isolated on either aspect of the equation.

8. True functions are able to assign one value for each input.

EXERCISES

37

9. The vertical line test informs that we can visually determine if the function is true. If any vertical line is crossing more than one spot on the graph, that function may not be a valid function.

10. Functions are able to be utilized as inputs into different functions.

11. Functions can be integrated to create new capabilities.

12. Inverse functions run at reverse. Given the outcome of a function, what actions do we have to do in order to return the value that was originally generated?

13. The inverse functions can be written in the inverse function f-1 (x) to indicate the reverse function of (x) to indicate the inverse function of (the x).

14. Inverse functions can be a mistake. the real function.

15. In a formula it is possible to create a new variable in order to symbolize a function.

16. Implementing variables in functions may simplify an operation and make it more straightforward to tackle.

17. When substituting an element for a function it is important to remember to get it replaced in the final.

18. Functions are able to include several independent variables.

19. Functions may include several dependent variables.

20. The logarithm function is reverse of exponentiation.

21. The natural logarithm is with Euler's number (Euler's number () as the basis, it is the "default"

logarithms for use for calculus.

22. Logarithms may be utilized to turn problems with exponentiation into multiplication problems.

Exercises

1. If you have f (x) is 3x + 5, what's the value of f (7)?

2. If you have f (B) = 2 + b, what's (9)? (9)?

3. If the formula f (x) = 2x + what's the value of f (3)?

4. If you find that f (n) = 2 n + 7 then what is the value of f (q +) 5. If the formula f (x) is x 2 + 3x How do you calculate the value of f (one + z)?

6. If the equation h(x) equals x4, then what would be h(3)?

9

7. If you have f (x and the number y) equals 3 x + 4y, what's the value of the number of (5 6)?

8. If the formula f (x) equals 5 x + x and the g(x) is 5x x - 9. What is the definition of f (g(x)) with regard to the x?

9. What is the reverse of the equation that f (x) = 1 + x?

10. What is the reverse of the formula that f (x) equals 5x?

11. What is the opposite of the equation of g(x) = x?

5

12. What is the opposite of the equation of g(x) = (1)3? 1)3?

13. What is the reverse of the formula of f (x) = the number ln(2)?

38

4. BASIC TOOLS: VARIABLES AND FUNCTIONS

14. Convert the implicit formula of x = 1 + x into an explicit function for the number x.

15. Convert the implicit formula x + 3y into an explicit expression of x.

16. Convert the implicit formula 3y + x = y into a formal function for 3y.

17. Find the equation below for the x number the following equation: Tan2(the x) + 6 tan(x) + 3 = 3 =. Make use of radians in calculations.

18. If you have f (x) is 3x, the g(x) is x 2, and the equation h(x) is equal to the number f (g(x)) What is the value of h(x) with respect to the x?

19. What was the significance of h(5) from the earlier task?

20. If the formula f (x) is sin(x) + 4 (x) = 4(x) is 2 cos(x) + 6, and the sum of h(x) is (x) + f (the x) plus the number g(x) what does the value of h(x) as a function of the x?

21. Let's consider the equation: y = 3 + x. Let's suppose there is a function of f (x) = 2x. The equation can be rewritten using the formula f (x).

22. Let's consider the equation: which is y = 6x + 5. We will assume that there is a function of f (x) = 3x. The equation can be rewritten using the term f (the x).

23. Let's consider the equation y = 2x. Let's suppose there is a function of f (x) = x + 1. The equation can be rewritten using the formula f (x).

Chapter 6: Basic Tools Graphs

The majority of people don't understand mathematics and assume that manipulating symbology is the main art of maths, but this isn't the case. For certain, manipulating symbols is extremely important and can help people think more concretely concerning abstract questions. The manipulation of symbols can help you analyze problems through their simplest, logical connections and assists you in solving issues that you not be able to envision what might happen.

But, math is, broadly speaking, about reasoning to change the elements we are familiar with in order to discover those things we don't have the knowledge of. It is sometimes presented to us as symbols or numbers, however other times, it appears through different methods for example, using the graph.

In many cases, there is no equation to describe a function. A lot of functions are only known via graphs. If you're in your business, it is possible that you have one of the sales graphs. It is unclear what equation determines the graph or even if there's an equation. The graph is just a graph. The equations can be approximated to it, but what are really able to see is the graph in itself.

In the field of science, one the primary duties is to look at the data in a certain way and determine what relationship has been established. One of the most effective ways to comprehend the nature of relationships is to comprehend how graphs from various functions appear similar to.

graphs are extremely important in calculus, particularly, since many concepts are easier to grasp when studying graphs.

This is designed to present a concise outline of graphing in math.

5.1

Thinking About Graphs

What exactly is graphs? The graph is simply a way to visually show the relationships between different the values. The most popular graphs are in two dimensions and show how two variables interact. Most often, the numbers are x and which is the value x (or the variable that is independent is) depicted by the horizontal axis while the y (or the dependent variable) appearing on the vertical axis. It isn't technically necessary however, following this standard makes it easier to understand the graph on a quick glance.

The graph's axes are basically number lines aligned so that they exist in their own dimension, and they meet at zero-points. They are referred to as Cartesian graphs because they utilize the concept of coordinate space that was invented by the philosopher and mathematician Renee Descartes. The concept revolutionized maths by offering a clear connection between geometry and algebra. By using this method 39

40

5. BASIC TOOLS: GRAPHS

It is possible to describe geometric forms by using equations. It may be intuitive to you today however, in 17th century, it was groundbreaking. 1

In the past, there have various methods of graphing, like the polar graph and polar graphs, which you've probably encountered. The main focus of this book lies about cartesian graphs. However, the techniques we develop could be applied to various graphing methods.

Chapter 7: Algebraic Function

We will look at an algebraic function that has the form , where the constant '' is that is referred to as a coefficient "'" is referred to as an exponent, and "'" is referred to as the variable. In this chapter, we will discover how to execute an mathematical procedure known as differentiation on the previous function. It can also be referred to as determining the derivative of this function.

The distinction must always be done according to a particular variable. For our example, the variable is , as it is said, we'll understand how to separate the function from .

Let's see how we can achieve it.

To separate the function from regards to , or more specifically phrases, find the derivation of this function compared to , perform a easy task, as described below.

"First multiply the constant co-efficient with the exponent, and then subtract the value of the exponent by 1".

Take a look at the formula below.

Formula:

Let's explore the formula more closely.

First see the L.H.S,

can be read as a distinction the two .

Now, see the R.H.S

As we mentioned earlier in the beginning, we first multiply the constant coefficient by the exponent . Then, we've diminished the value of the exponent by one . All you'll need to accomplish.

Let's clarify everything with an illustration.

Example 1: Imagine that you're required to calculate the derivation of a function (w.r.t the word "HENCEFORTH" in this book, Regard to WILL BE SCREEN in the form of w.r.t)

How do we do it? Easy enough.

Then, you must compare the equation to .

We can easily say you are in a different position and .

Let's apply the formula that we learned, and then insert the numbers correctly.

We are familiar with the formula that allows differentiating,

Thus, a derivative of can be written as .

Inserting and within the R.H.S in the above formula,

We get,

This is it, you've succeeded in separating functions w.r.t and you know the answer .

2. Find the derivation of w.r.t .

To begin, we must compare it with the formula. We can discern that . After that, put the appropriate numbers into the formula.

We get,

We have a solution.

3. Find the derivative with respect to .

What can you find there? It's simple, and instead of the usual mathematicians the term used in this example can be used to describe the situation . Replace it in the formula and repeat exactly the same procedure.

The formula now changes to

Let us consider the above function and evaluate it against . It is easy to say "What is the function that it's equivalent to? It is important to know why? Since it is the same . Hence and .

Notice: If there is the coefficient of exponent cannot be visible, then the figure is interpreted as 1.

Therefore,

It is now time to master the fundamentals of carrying the differentiation of algebraic functions efficiently. You should take your time to learn again and in no time you'll be able to enjoy discovering the derivatives of functions.

Let us move ahead. Some times, the algebraic functions or polynomials appear fascinating and distinct. Here are some tips to get it into the right format.

1. If you notice a field on the form, you can rewrite it to be . Then you will be able to see

it, and you are able to easily contrast it with the format that provides and .

The idea is that if you can see the variable as denominator of the equation, then bring it up to the numerator by altering the form of the exponent. Examples:, etc.

2. If you're being requested to separate from the function, then you need to write the function using a different name, and a hint is that if you find an inverse square root, then compose it with an exponent the same way cube root could be written in exponent form and on.

Check out the useful formulas for exponents.

,

,

3. If you're required to distinguish 10 from 10, what do they signify? What is the correct answer? There isn't any factor in any way. This means that 10 may be interpreted as contrast with .

We get ,

Thus, the formula provides the RHS

As zero is multiplied and multiplied, the end result is zero. Thus, the result is that the "derivative of a constant" is always zero.

Example : , ,

The functions all are constants in the examples above.

4. If you're being called upon to make distinctions, initially it is clear that the functions or polynomials directly are not the same as . What you could do is separate each of the terms and then summarise your final results. Meaning,

Write in the form of

Then you are able to determine the distinction of every term separately, since each word is separate by its own shape .

The ultimate answer is

"If the polynomial has many terms, find the derivative of each term individually and then add them together"

Let's take another instance to get a clearer understanding.

Differentiate:

Compare that with now Now, compare with

Constant

The ultimate answer is

In the present the above is all that you must remember when making the distinction between an algebraic function and an algebraic polynomial. It is time to build your foundation and build our practice through completing some further exercises.

Exercise

Find the derivative of every of the above functions in relation to the specified variable. Try them on your own initially, then examine the solution to verify the answers.

Exercise 1:

Solution: Compare and resolve as per the formula , and compare it to the formula ,

By applying the formula, you get

Exercise 2:

Solution: Here, the variable is . Therefore, we must make use of the formula .

Compare, and with . Like we said earlier, if the variables is part of the denominator of the equation, change the variable in the numerator. It can be written as the comparison tells us that

Incorporate the value in the formula

Exercise 3:

Solution: Rewrite the formula the formula as Comparing with and . By applying the formula, and adding appropriate values

Exercise 4:

Solution: Compare it with ,

By applying the formula, and then inserting the numbers,

Exercise 5:

Solution: Examine with and

When we insert these values into the formula, we'll get

Exercise 6:

Solution: First, it is important to be aware that the variable is . Thus, the formula to use in this case can be described as follows:

Then, to see how it compares with the other start by writing it in the form of

If we look at a comparison with when compared with or . After applying the formula and adding the numbers we receive,

Chapter 8: The Chain Rule

In the last chapter, we were taught the ways to distinguish algebraic functions in the form , and also learned to put this function into the previous structure before separating it when the formula is given in an alternative format. So, what exactly does this chapter cover? Consider a question like this

You will be able to find

Examine this formula a little more closely and then test it against and to determine the value of and to utilize to the formula.

Then, you'll see that it isn't simple. It's a laborious task to multiply fifteen times, then search for the meanings of and the many concepts that result. If yes, what is the best way to distinguish between method or figure out the w.r.t

The most simple and efficient method for doing this. It is called"chain rule.. What do we do with the chain rule? Take a look at the role of this rule and the goal as we all know, is to determine the the derivative w.r.t .

The first step is to convert it into a function for some different variable, for example . It is accomplished by simply changing the old function using a letters .

i.e. Write

The initial function is now an expression of (here is the variable) and we are aware that it is dependent on (here can be the variable). The chain then goes as follows: the function provided is to discover the answer or

Writers compose . Thus, the function changes to and . We will call it an outer function, for the sake of simplicity and call it the inner function. The chain rule states that or is equal to the derivative product of the function that is outside and the inner function. It is as follows,

or

Example 1: Let's be clear about this issue by setting into consideration this scenario. Differentiate

Then write the first line or

The function is now called the outer function. Let's call it the inner function.

Formula says,

and

.

Therefore, the ultimate answer is

Example 2: Find

The original purpose is

We must find out the derivative of .

Chain rule applies to us. We first write . The original function is now an outer function, also known as the inner function that has an variable . The inner function is referred to as that has the variable . Formula says,

and

So, the answer in accordance with the chain rule would be,

Or

3. Find the derivative of the function w.r.t

We've been asked to learn more about the situation whether or not we can.

First, let us rewrite it in the form of

So . Let or . Thus, the function is"[Outer Function] as well as [Inner Function[Inner Function]. The chain rule formula states,

The final solution is

Or

Then you'll have an concept of how to use the chain rule in order to differentiate functions. We can try a few more tests to help make the idea more solid. Do your best to figure it out by yourself, first. Then check the solution with the help of a colleague.

Exercise

Find the derivative of function given with respect to the variables they represent.

Exercise 1:

Solution: Let us must find out

Let or . The function that was originally created is now"[Outer Function] as well as [Inner FunctionInner Function. Chain rule says,

Let's solve each problem in turn, then divide by

Therefore, the ultimate answer is

Exercise 2:

Solution: Let's must find out

If you want to, the original role is changed to and

As per chain rule

So, the ultimate answer is

Apply the principles into the values

Exercise 3:

Solution: We have to know

It is possible to change the initial purpose is changed to , and the function inside

Chain rule is applied to chain,

Let's tackle each problem individually, and finally divide by

The final solution is

Exercise 4:

Solution: Here , Let

and

Use chain rule.

Let's define each one in a separate way, then we can multiply them.

Chapter 9: Product Rule

We will begin with this brand new chapter by discussing one of the key ways to calculate, which is known as the product rule. The rule can help us distinguish a particular function that is defined in the form of a product. Take, for example, this problem

Differentiate

You can clearly see that this function appears as a product of two roles. i.e. it is in the format

Both and function on the identical variable .

Does anyone know how to separate these functions, without expanding and multiplying the two terms of the product, the answer is a resounding yes. This is known as the product rule.

The formula of the product rule says

The first function is multiplied by the derivative.

Example 1: Let's apply the same formula that we used before, but differentiate.

It is important to note that the equation follows the formula . The product of two functions and two variables .

Here and

We can apply the rule of product differentiation,

It is a form,

Let's complete the calculation.

Therefore, the ultimate answer is

This can further be reduced to

Example 2: Differentiate

The form is . This is the First Function as well as a The second function.

Follow chain rule.

This can be further streamlined to be referred to as

Example 3:

This issue is in the form that consists of - Ist Function, and II Function.

We can distinguish using the principle of production

Exercises

Exercise 1:

Solution: Follow the rule of product, which is the format

Exercise 2:

Solution: Using the rule of product

This can be further simplified If you want to.

Exercise 3:

Solution: and

Exercise 4:

Solution: Applying the rule of product,

--- Equation 3.1

It is possible that you have spotted that the final differential of the equation above i.e. it should be solved with the chain rule, let or and

Substitute in Equation 3.1

We get

Exercise 5:

Solution: Applying the rule of product here, and

--- Equation 3.2

Ist Part IInd Part

Chain rule is used to calculate differential in the Ist portion.

Let , and

Use chain rule to determine differential in the IInd Part

Let , and

Substitute the back of Equation 3.2

If needed, simplify it.

Chapter 10: The Quotient Rule

In the last chapter we learned how to distinguish functions, that are by way of a the product of two functions by using the rule of product differentiating.

In this section, we'll be taught how to recognize the function that is composed of the numerator and denominator. i.e., . As an example, how can you distinguish between a particular function that appears similar to the one above? The answer lies in quotient rule that we'll learn in the near future.

This is because the purpose that will be differentiated can be described in the form . In the above example you will be able to see the difference.

and

It is referred to Numerator Function as the denominator function, for convenience sake.

Then, what exactly is the meaning of the quotient rule?

The formula for the quotient rule is,

Simple mathematical concepts

We will make it easy by using an illustration

Example 1: Differentiate

You can clearly see the function of the numerator and denominator function . According to the quotient rule formula,

We should fill in every word within the R.H.S and figure out the solution.

Incorporate all of these responses in the R.H.S of the quotient rule formula.

This is all we need to do, we've completed the differentiation according to the formula. Now you can simplify the process and type.

The final solution is

Example 2: Find

The numerator function, as well as the denominator function are . We can apply the formula for quotients,

First, denominator function

Numerator function derivative

Function Numerator

Function derivative of the denominator

Function of the square of denominator

We can plug in the terms we have derived in the formula for quotient rule

That's our suggestion This is the best way to go, but you can reduce it more if you wish.

You have learned how to distinguish a particular function that is as a quotient with the help of the quotient principle of distinction. Try a couple of further exercises to help make our skills more effective.

Exercise

Exercise 1:

Solution: Let's use the formula of the quotient rule

Here, and

Exercise 2:

Solution: You can use the chain rule or the quote rule to address this issue. This is a case where a constant.

Chapter 11: Trigonometric Functions

In the preceding section, we learned how to distinguish different kinds of polynomials and algebraic functions by applying the appropriate rules for the particular circumstance, like chain rule product rule the quotient rule, and sometimes a combinations of these rules. But, these methods and rules aren't limited to functions that are algebraic and are also applicable to various other types of functions including trigonometrics, logarithmics and exponential, etc.

In this section, we'll study and apply differentiation whenever trigonometric operations are in play. We are all aware of trigonometric identities as well as relationships such as

For you to be able differentiate trigonometric functions, it is necessary to learn and memorize the following Trigonometric function derivatives listed in the table below.

"Write it a few times, and memorise these derivatives of basic trigonometric functions"

Let's take a look at some examples of how trigonometric function can be distinguished.

Example 1:

This is where the trigonometric functions are . The expression does not appear explicitly in the format of the trigonometric basic function, neither is it specifically within the structure . Therefore, we use chain rule

Let ,

In this case, the function that was originally intended remember that we identified this as the outer function. The inner function is also called. Chain rule is applied,

Example 2: Let's look at another instance involving the rules of manufacture.

Chapter 12: Exponential And Logarithmic Functions

The differential of exponential functions

The fundamental exponential function in which the underlying constant, and is also called Euler number. This approximates, and is an variable. In this chapter, we'll be taught how to calculate the exponential function's derivatives as well as the formula to calculate the formula for a derivative from an exponential function . Below is the formula,

You can clearly see that both constants were multiplied. The exponent remains the way it was. We will look at some basic examples to grasp this. Consider an equation . When you contrast it with . We know that, and .

We can now use these values to the formula

If so, then the answer is

The answer seems a bit odd, isn't it?

If you separate the functions w.r.t it, the answer will be the same . Let's take a look at some other instances

Example 1: Differentiate

Check out the comparison Here . Use the formula

Example 2:

Based on the comparison to the one we have . By applying the formula

Example 3:

The first thing to note is it is the variables . Therefore, the formula changes.

Another thing to be aware of is that it isn't specifically in the shape . Therefore, let's use this chain principle. Then, the task changes to ,

The basic concepts of a differentiation exponential function. We can now tackle a couple of more exercises to help us improve our skills.

Exercises

Exercise 1:

Solution: If you take this as a function, it's at the end of the line, and

After that, applying the quotient rule.

Utilizing the property of exponents

Or Method II

It is possible to rewrite the function in its original form.

You can now differentiate the two with the formula that you use

It is evident that both approaches give the same result.

Exercise 2:

Answer: Now, think about this particular function . We can divide the differential in two ways:

We'll tackle one at the time

First,

Second,

We can apply the chains rule because the function above isn't necessarily as a result of

"Let and" inner function

The function -- The outer function

Third,

Add all the terms now,

Exercise 3:

Solution This function comes from the form of where and

Application of the product rule

Exercise 4:

Solution: The function is , we need apply the chain rule in order to resolve this.

Let and --Inner function.

Then -outer functionality.

Substitute for

Exercise 5:

Solution: The purpose is in the shape . Here and . We can employ the quotient rule.

--- Equation 6.1

In the equation above it is necessary to determine the solution for . Apply the chain rule. So that the answer is us then

Substituting in the Equation 6.1, we get

Logarithmic logarithmic function differentiation:

Consider a number like 10,000. It is well-known that it is also possible to write in the form of . If the logarithm of 10.10 is four. It means the ratio of a base number is the exponent of that base, which results in . In the example we have provided it is the base and the number of exponents over base is 4. So, the logarithm for 10,000 times 10 equals 4.

Written as

The way we write as a logarithmic operation. If we take 10 as the base number, they are known as decimal logarithms. Sometimes, they are also referred to as the common logarithm. There is possibility of using an Euler numeral as the base. These logarithms are referred to by the name of Natural logarithms. They are widely used in the fields of mathematics, physics and engineering areas.

Natural logarithm functions are depicted as follows: here, it's used as the basis.

In this article, we will discover the ways to recognize logarithmic function. Prior to that, take an eye on the following table which contains the properties of the natural logarithmic function.

1. is often described in the form of

2.

3.

4.

5.

6.

7.

We'll now come to the principal aspect of this article, the distinction between natural logarithmic function.

Let's look at a few cases of differentiating operations involving natural logarithms.

Example 1:

Solution: This is in the form of where and . Utilizing the product rule

Is the only way to find out.

Examples 2: this is a function that involves the quotient rule. Here

By applying the quotient rule

Chapter 13: Derivative At A Point

Think about a function. Now we are aware that is a variable within the function, and is usually portrayed by a value to a certain amount of . If, for instance you insert it into the function mentioned above, it is transformed into

So, the importance of the function . or when . You can also input any value or, you will get an appropriate value to the function you are using. The information above. It is not necessary to provide to explain what a purpose is.

The question that needs to be dealt with in this chapter concerns what can we do to determine what the worth of the derivative of a given function at a particular point.

Take the same equation into consideration as in the case

What's the worth of the function's derivative for ?

In order to understand and resolve the issue, we must recognize that the issue here is not about finding the value of the function [43]. The question instead is to determine what the significance of the derivative .

Then we can discover the derivative of this function. This is

Then we can observe that the derivative of this function is determine the value of the derivatives, simply change the value by the value of . Thus, the amount of the derivative of the function is .

Such problems can be quite significant in applied maths and physics. Along with determining the derivative of an equation, the idea of determining the worth of the derivative for the given level important.

Let's look at several more cases.

1. Find the worth of the derivative at .

The first step is to find the component of the function whose derivative can be found by calculating the derivative of the function.

It is necessary to determine the worth of

2. Find the amount of the derivative

The first step is to determine the derivative from the formula w.r.t

Then substitute the derivative above

Take note: If you're confused as to what we did to get here you are, we've applied the chain rule. Let , and

Chapter 14: Second Order Derivatives

Consider a function , and if we need to determine the derivation of this function w.r.t the given function, we apply the basics of differentiation and find out the following answer:

So, that the derivation of the function is . Let's go one step forward by separating this answer w.r.t . That is, let us be able to distinguish w.r.t . We can then response as:

What did we do to achieve this? We've reconstructed the first function twice in succession. This is referred to as locating the second derivative, or Double derivatives of the formula. The example we have given shows the function that was originally used was

The first derivative is known as.

Also, it is referred to as a differentiation or a second derivative of, and it is typically known as .

refers to the fact that is nothing more than the first derivative

Second derivatives is a concept is a broad concept that has a variety of uses in physics as well as engineering, when you have to improve a particular function value as well as find its maximal and minimal values, that is taught in the higher courses.

Let's take a few more instances to grasp the idea in a clear way.

Example 1: Find if

The initial variant w.r.t is

Second derivative w.r.t is

Example 2: Find

The purpose of the function is

The very first derivative w.r.t

A second derivative w.r.t

Is the answer to this question.

Example 3:

The function that is given is

First derivative,

The second derivative

Exercise

Exercise 1: Find

Solution: The function has the form . Where and . Find out the initial derivative made using the the rule of product,

Then let's differentiate this answer to get another derivative. The part to be differentiated is in the form

The rule of product application applies to both words

Exercise 2:

Solution: The task is in the shape

The primary variant is

Then, by dividing once more to get the second derivative,

Exercise 3:

Solution: Let

The initial variant is

[Hint: Use chain rule]

We will now separate the function above in order to obtain the second derivative.

Exercise 4:

Solution: Let

First derivative [Hint: Chain rule]

Second derivative

Exercise 5:

Solution:

First derivative,

Another time to divide the second derivative

INTEGRATION

Chapter 15: Integration

Polynomials and algebraic functions

In the first section of the book, we learned how to carry out the mathematical procedure of differentiation. It is a process that allows for the different kinds of functions. We also discovered a range of different rules that can be used for performing the same in various conditions. In this section of the book, we'll discover how to execute a second mathematical operation known as integration that is is a vast subject that has many applications in a wide range of areas. But, we'll cover only a handful of aspects in the book considering the point of view from the beginner.So we will start.

Take a look at a function, or there is a constant coefficient. is the exponent, and the variable. Which operation can be performed called integration of this function w.r.t . What is the integral of this function w.r.t ? The formula or method is given in the following table.

Let's look at the formula in both L.H.S as well as R.H.S with greater care and be able to comprehend the formula fully.

Then let's take a look at the L.H.S

In the beginning, you will see an extended symbol that signifies integration just similar to the symbol that signifies differentiation.

You then have the function to be integrated, and finally you will have a symbol to indicate that integration is completed w.r.t to the variables .

Let's take a peek at the RHS

Which can be the constant coefficient for this function, which stays as is.The strength of the function is increased by i.e. and the entire term gets divided by .

There is a third constant that is also known as The constant of Integration. It is customary to include this symbol in the conclusion [Not indefinite Integrals, but we'll be learning very soon] which indicates that you've incorporated this constant to your answer. The importance of this constant will not be fully discussed within this text. If you are studying calculus in upper classes, and you learn concepts like antiderivatives, boundary conditions and so on, the importance of this constant should be obvious for you. Let us just add the constant in every solution is obtained after the integration.

Like always, let's be clear about the process by taking some of the instances.

Example 1: Incorporate the function w.r.t

We can integrate this function w.r.t in accordance with the formula

Let's create the LHS

By comparing the formula to ours, we know that

We can substitute the right values into the RHS to find the solution.

This is all we need to say, we've achieved the integration w.r.t and we have the solution

Written as,

Remember, the continuous we include at the bottom of each answer in this textbook. (Except for definite integrals).

Example 2:

The variable here is in place of . We can rewrite the function to read . Compare

With

Here and . Use the formula

Example 3:

First, let us rewrite . Then let's compare to .We not only note this and .

Utilize the formula

Example 4: Integrate

Check out the functionality.

It is a form.

As we learned in differentiation that, in the case of a polynomial term could be separated and added to, the same way when integrating each term can be integrated in a separate way and then added.

If so, then

Consider the following example

One constant at the end of each is enough.

Note 1: There's an exception to the application of formulas it is . is not equivalent to [. In this case, as we observe from the formula R.H.S is . This gives us [infinity] this is not a true answer.

Note 2. Integration is the opposite of operation of differentiation. Example: If it is, then .

Let's solve a couple exercises to help us understand the concept better.

Exercises

Exercise 1: Integrate

Solution:

Combine each of the terms and compare with the formula

Exercise 2:

Solution: Let

Exercise 3: Integrate

Solution:

Exercise 4:

Solution:

--- Equation 9.1

We must integrate every word in turn, one at a time.

In this case, the term integrating is , or in other words what's the integral of an infinite constant. It's not zero as it pertains to the case of differentiation.

The first step is to write it the text and then check it against .

It is clear now is

By applying the formula, we will get

A constant is integral in the case of the constant.

Let's write the ultimate answer to our question by replacing the words in Equation 9.1

Chapter 16: Definite Integrals

In the preceding section, we were taught the ways to incorporate different kinds of function w.r.t the variable being studied. The information we have learned up to this point can be described as indefinite integrals. This are integrals that have no limits. In this chapter, we'll discover how to analyze an indefinite integral. This is also a great tool for engineering, physics and relevant disciplines.

Let's look at an example of a role.

If we're simply being asked to combine the function w.r.t it is now it's a very easy task for us. Apply the formula then compare the two terms to find the answer.

Here's the term we use to describe an undetermined integral.

So what exactly is an essential?

A good example for the definitive integral questions would look such as, analyze the integral of the lower limit the upper limit

As you can observe in the previous question the definite integral must be assessed in two different limits.

This is considered to be the lowest limit while is the top limit.

So, how can we tackle this problem.

The formula for solving an integral that is definite is provided in the following table.

Let's examine the formula more closely and then use an example in order to make the process crystal easy to understand.

The L.H.S is written in the form of, which indicates that the function should fit within limitations . The word is placed on the lower edge of the integral symbol. The other side is located on the upper part of the symbol . It is referred to by the name of lower limit, and is also known as the upper limit.

Focus on the R.H.S the text reads. This implies that once you incorporate the functions w.r.t you will find an answer, which has a different function . It is called . Then, substitute the limit that is higher than what it provides you with the lower limit which provides you with . Finally, subtract

It will then give you the ultimate answer.

Let's take this example that we discussed at earlier

Example 1:

It is evident from this query that the function given represents the limit lower, while is the higher limit. Then, firstly, examine the integral in a similar way as you would normally.

Do not now add the constant, as it isn't an integral indefinite[Now do not add the constant as this is not an indefinite integral

You can now see the result obtained following integration again is a result of . It is referred to the following:

Use the formula

This is now achieved by substituting into, which results in

This is accomplished by substituting it in this way, giving

The final solution can be achieved through substitution

This is the final solution.

Let's look at a few additional examples to help make the idea simpler.

Example 2:

Here's the task to be incorporated. The lower limit is, and is the top limit. Apply the formula.

Let's first find the integral in the normal way

We can see now that

Replace the upper limit

Substitute for the lower limit

The answer to the final question is

Example 3:

The purpose is . Below are the limits of lower, and is the higher limit. Utilize the formula

First

Set the upper limit

Use the lower limit

The answer is the final one

Example 4:

Here

Upper limit

Lower limit

Final decision

Let us now practice a few more exercise questions.

Chapter 17: Rules Of Domain Of A Function

1.) In the event that f(x) is equal to an equation that is a quadratic, linear or cubic function without a fractions or square roots typically, the domain comprises the whole range of real numbers.

Important The real numbers are negative and positive numbers. The majority of integers are real however, not all numbers are actually integers. Real numbers can also have an element of fractions, however the integer is not able to possess a fractional part.

E.g. 1.) The area of the equation f(x) is 2X-7. It encompasses the entire range of real numbers, all the way from negative infinity until positive infinity. In interval notation, it is expressed by (+,).

2.) The area of the equation f(x) is X2 plus 3X 5 includes the entire range of real numbers, all the way from negative all the way to infinity positive. In interval notation it's expressed in the form (+,).

3.) The area of the equation f(x) equals 2X3-5X2+7X-3 comprises the entire range of real numbers, all the way from negative infinity until positive infinity. In interval notation it's shown in the form of (+,).

2.) If f(x) is an rational function, such as fractions, the domain will be generally any real number which isn't less than 0 (0.).

E.g.1) The area of the equation f(x) is any number in the real world, except for 2. X 2 since 2x = zero. In interval notation, it is the amount of money donated in the form of (2)) (-, 2) (2,).

2.) The area of the equation f(x) = can be any number with real numbers, the only exceptions being root of which is 4 5, and 4. X 4, X 5. In interval notation it's expressed by (-4,) U (4 5,) U (5,).

Rule 3) If f(x) is equal to its square root a formula with no fraction The domain usually larger or equals. (X>=0)

E.g.1) The range of the formula f(x) = could be any number that is real from four to infinite. In interval notation, it is expressed in the form of (4,).

TYPES OF FUNCTION

1.) COMPOSITE Function: The concept behind a composite function is the substitution of the X.

EXAMPLES:

1) Given F = X2, g = X+1, evaluate F o g.

SOLUTION

Input g into f. F O G = (X+1)2

2) Given g = 2X+1, h = 3X+2, f = 2X. Find F o h o g.

SOLUTION

Find h o g first = 3(2X+1) + 2 = 6X+3+2 = 6X+5

Therefore, F o h o g = 2(6X+5) = 12X+10.

2.) INVERSE of a Function The purpose of the an inverse function is the ability to switch or interchange between X and the Y.

Please note that F(x) does not mean the exact concept as the same thing as Y. F(x) is Y.

EXAMPLES:

1) If f(x) is 5X-3 3. F-1(X) is What?

SOLUTION

F(x) = Y. Therefore, Y = 5X - 3. Exchange Y for the number X.

X = 5Y - 3. Add Y as the modification of formula for the subject.

X+3 = 5Y. Dividing both sides by 5.

= Y =

F-1(X) =

2.) In the case that f(x) is 3+4X2, how do you find the reverse of this function (F-1(X))?

SOLUTION

F(x) = Y. Therefore, Y = 3+4X2. Exchange Y for an X.

X = 3+4Y2. Change Y to the formula for the subject.

X-3 = 4Y2. Divide both sides by 4.

= Y2 = square on both sides.

= Y =

Therefore, Y = . F-1(X) = .

3.) ABSOLUTE VALUE FUNCTION Absolute value functions for all XeR is described as

|X| = f(x) =

{x if x 0

{ if x = 0

{-x if x 0

The program assigns positive numbers of a particular number to that number, and attributes 0 to zero.

EXAMPLES:

|4| = 4, |0| = 0, |-7| = 7.

A word of caution: Absolute value keeps positive numbers and zero just as it is and also transforms negative numbers into positive.

ATTENTION: TO LEARN HOW TO SOLVE BASIC FUNCTION QUESTIONS WITH SCIENTIFIC CALCULATOR VERY FAST IN 5 SECONDS, ENSURE TO CHECK OUR YOUTUBE CHANNAL. CHANNEL NAME IS EAGLESCLASS

PRACTICE QUESTIONS

1.) The area of the formula f(x) equals) (0,

2.) The result of the function is g (x) =

A) (144b) 49c) 60) None

3) For all x e R, f (x) =

{If the x value is 0 what kind of function is (x)?

{x, if x

a) Heaviside function b) Signum function c) Absolute value function d) Power function

4.) The reverse of the equation (f) (x) is 8x + 3..) 3x 8 (b) 3x + 8. c)

5.) What common domains would the identity and the absolute value function the same? A) The real set of numbers. b) The number of positive numbers) The number of positive integers

D) The rational set of numbers

6.) Let the following be a:

a) (x + 1) b) (x +

7.) Which one of the following functions can be described as one of the following:) f(x) equals the x) (x) = g (x) =

A) I, II I, iii, and iv) II, I and Iv C) II only) I, iii, and iv alone

8.) The reverse of the equation f (t) =

A) F-1 (y) =) (y) = b) (y) is c) F-1 (y) equals d) F-1 (y) isn't defined.

ANSWERS

1) B.

2) D. g (x) = [

= [

3) C.

4) C. F (x) = 8x + 3

Let y = 8x + 3

Y - 3 = 8x

X =

1. (x) is x (y) =

5) C.

6) B. f (x) =

F o g = f (g (x)) = f (x+1) = (x +

7) B.

8) B. Let f (t) = y

y =

1. y (1 + 1 +) is 1 + t

1. + t

The y -1 value is yt + t

1. y + 1 = (y + 1)t

Chapter 18: Limit Tending To A Real Number

Important that real numbers are positive and negative numbers, which could have a fractional component. Methods of calculating limits that lead towards a true number can be accomplished the following: 1.) Through direct substitution, 2)) Through factorization, and 3) by the L'Hopital rule.

DIRECT SUBSTITUTION METHOD

EXAMPLES:

1) Evaluate

SOLUTION

= 2(2) - 7 = -3

2) Evaluate (3

SOLUTION

(3 = 3(0)2+2(0) +1 = 1

FACTORIZATION METHOD

Utilize the rule of factorization prior to using direct substitution to determine the upper limit.

EXAMPLES:

1) Evaluate

SOLUTION

= = = 2+2 = 4

2)

SOLUTION

= = = = = =

L' HOPITAL RULE

It is the process of determining the derivatives of the numerator as well as the denominator for a rational function, and then applying direct substitution to identify the limit.

EXAMPLES:

1) Evaluate

SOLUTION

= = = = 2(2) = 4

2) Evaluate

SOLUTION

= = = = =

3) Evaluate

SOLUTION

CONTINUOUS AND DISCONTINUOUS LIMIT

CONTINUOUS LIMIT: Limit can be defined as continuous if it is X's value at the location is (defined).

EXAMPLE:

The expression is with 4. The limit can be said to be continuous due to the value is 1.909 (a specified number) thus the function has been established.

Limits are described as discontinuous when it is determined that the level of X in a location isn't there (undefined).

EXAMPLE:

In the equation with X = the boundary can be described as non-linear, since = is not defined.

Types of DISCONTINUOUS LIMIT include: 1) Essential discontinuity 2.) the ability to remove discontinuity.

Essential discontinuity is a result of a limit that exists, and it is a removable discontinuity that occurs when the limits exist but do exist.

ATTENTION: TO LEARN HOW TO SOLVE LIMITS WITH SCIENTIFIC CALCULATOR VERY FAST IN 5 SECONDS, ENSURE TO CHECK OUR YOUTUBE CHANNAL. CHANNEL NAME IS EAGLESCLASS

PRACTICE QUESTIONS

1.) Make [x] the biggest integer that is less or equivalent to x. Find

A) 3b) 2 3) Is not present d)

2.) Assess the following:) 8.) 4 C) 7 d) 0) 7

3.) Locate the points that are discontinuous in the function S(x) =

a) b) c) d)

4.) If the function f is continuous with x equal to a the answer is (a) f(a) (b)

C) doesn't exist) f(a) doesn't exist

5.) Evaluation of yields)

6) Evaluating

7.) Find the upper limit of the equation (g) = as (x) = x

8) The

9) The answer is) 1 2.) (d)

10.) Examine the following:) 1 1.) (c)

11.) Solution:) 2 (b)

12) Find the

D) It is not there

ANSWERS

1) A.

2) D.

3) D.

4) B.

5) B.

=

6) A.

Consider the ratio between most important coefficients. = =

7) D.

We got zero through substituting, so we take into account prior to replacing

8) D.

9) D.

10) A. . Direct substitution: place the number x equals

11.) C. Use L' Hopital rule i.e use the product of the numerator as well as denominator prior to using direct substitution.

12) A.

DIFFERENTIATION (DERIVATIVES)

GENERAL DIFFERENTIATION FORMULA:

Note If you separate between a constant(real numbers) this gives you zero as an amount.

PATTERN OF DIFFERENTIATION

FUNCTION DERIVATIVES

$y = 3x$

$Y = 2$

The formula for y is 3x plus 8

TYPES OF DERIVATIVES (DEFFERENTIATION):

1.) The first derivative, which is called Y and F|

2.) A second derivative, which is called or F|

3.) Third derivative called also F|

RULES OF DIFFERENTIATION

The rules of differentiation are:

1.) Sum and Difference rule

2) Product rule

3) Quotient rule

4.) The chain rule (function of the function)

SUM RULE AND DIFFERENCE RULE

Chapter 19: Derivatives Of Trigonometric Functions

NOTE:

I

EXAMPLES:

1) If y = 3 , find

SOLUTION

Application of the rule of product

Let u = and v = 3

3 = y , = or

2) Given y =

SOLUTION

Application of the rule of product

Let u =

"y" = "Y" ,

DERIVATIVE OF TRANSCENDENTAL FUNCTIONS

Natural logarithm function

the y value is in X.

EXAMPLES:

1) Given y = In 2x, find

SOLUTION

The 2x is y,

2) Y = In

SOLUTION

Y = In

3) Given y = , find

SOLUTION

y =,

4) y =

SOLUTION

"Y" =

IMPLICIT DIFFERENTIATION

Note: The concept behind implicit differentiation is that it allows for changes to the subject formula.

EXAMPLE:

1) Find

SOLUTION

, =

=

APPLICATION OF DIFFERENTIATION

TANGENT AND NORMAL TO A CURVE:

At (X,Y):

The slope of the tangent is equal to the slope curve

m1 =

Equation of tangent = + y1 = m1 (X * 1.)

Normal m2 slope = equals

Equation of Normal = y-y1 = m2 (X 1)

EXAMPLES:

1.) Determine the equation for the tangent as well as normal to the curve. y =

Chapter 20: Velocity And Acceleration

EXAMPLES:

1) If S(t) = -16

SOLUTION

S(t) = -16

32(0) + 100 = 100ms-1

2.) Following t seconds the object that is moving has traveled one S meters, where S = 4 when is the the object temporarily still?

SOLUTION

S = 4

V(t)

V(2) = 8(2) - 4 = 12 ms-1

V(3) = 8(3) - 4 = 20 ms-1

In the case of a stationary object, V(t) = 0

8t - 4 = 0

8t = 4

Therefore t = = 0.5s

3.) Distance S measured in metres traveled by particle S = 20t + 12, find the acceleration following 4 seconds. The moment when acceleration will be at zero?

SOLUTION

S = 20t + 12

V(t) =

a(t) = 24 - 12t

a(4) = 24 - 12(4) = -24 ms-2

If a(t) equals 0,

24 - 12t = 0

24 = 12t

T = + 2s

ATTENTION: TO LEARN HOW TO SOLVE DERIVATIVES OR DIFFERENTIATION WITH SCIENTIFIC CALCULATOR VERY FAST IN 5 SECONDS, ENSURE TO CHECK OUR YOUTUBE CHANNAL. CHANNEL NAME IS EAGLESCLASS

PRACTICE QUESTIONS

1.) If the y value is (4x +3), find the number of x = 2) 500 B) 2500) 5200) 40000

2.) Based on the fact the fact that 4

Chapter 21: Basic Standard Integral

1. POWER RULE:

It is due to integration being opposite of differentiation. Although differentiation decreases the strength of X the power of X.

EXAMPLE:

Evaluate

SOLUTION

Utilize the power rule

=

2. RECIPROCAL RULE:

EXAMPLE:

1. Evaluate

SOLUTION

3. INTEGRAL OF EXPONENTIAL FUNCTIONS:

EXAMPLES:

1. Evaluate

SOLUTION

2. Evaluate

SOLUTION

4.   INTEGRAL   OF   TRIGONOMETRIC
FUNCTIONS:

EXAMPLES:

1. Find

SOLUTION

2. Find

SOLUTION

GENERAL EXAMPLES

1. Evaluate (2

SOLUTION

(2 2 + =

1. Find

SOLUTION

In the case of an integral We can see that the result of the numerator is exactly the equivalent to the one from dividing the denominator.

F(x) =

2. Find

SOLUTION

Integral form of this kind is able to be broken down into a partial fractions and then incorporated for the purpose of obtaining

DEFINITE INTEGRAL

Definite integrals are integrals which are characterized by their upper and lower limits. They are given by . Definite integrals can be useful for finding curves that define areas and determining volumes of solids as well as many other uses.

EXAMPLES:

1. Evaluate

SOLUTION

In the first place, we are aware we are aware of

2. Evaluate

SOLUTION

Start by taking the integral that is indefinite.

APPLICATION OF INTEGRATION

Integration is a subfield of calculus which has applications across a variety of fields like engineering, mathematics, physics as well as management and social sciences. In maths, it's employed to determine the surface under a curve as well as the volume of revolution area of revolution, length of an arc, as well as probability density. In the field of physics, it's used to calculate the movement of a particle in the time that is t, the work performed by the force of a movement such as the moment of inertial.

www.ingramcontent.com/pod-product-compliance
Lightning Source LLC
Chambersburg PA
CBHW071218210326
41597CB00016B/1858